1 MONTH OF
FREE
READING

at

www.ForgottenBooks.com

By purchasing this book you are eligible for one month membership to ForgottenBooks.com, giving you unlimited access to our entire collection of over 1,000,000 titles via our web site and mobile apps.

To claim your free month visit:
www.forgottenbooks.com/free896443

ISBN 978-0-266-83456-4
PIBN 10896443

FOURTH SALE OF FINE
EARLY AMERICAN FURNITURE

GATHERED BY

JACOB MARGOLIS

CABINET MAKER

OF NEW YORK CITY

INCLUDING AN EXTREMELY RARE MAPLE DAY-BED
OF THE 17TH CENTURY; AN IMPORTANT QUEEN
ANNE MAPLE HIGHBOY; A FINE SPANISH-FOOT
WALNUT LOWBOY, ABOUT 1730; VERY UNUSUAL
SPANISH-FOOT AND BALUSTER-BACK ARMCHAIRS
AN EXCEPTIONALLY FINE SERIES OF LOWBOYS
AND HIGHBOYS IN CURLY MAPLE, CHERRY WOOD
ETC.; A SET OF TEN CHIPPENDALE MAHOGANY
SIDE CHAIRS; AN IMPORTANT HEPPELWHITE
SECRETARY-DESK AND AN INLAID HEPPELWHITE
SIDEBOARD; WITH EXCELLENT SPECIMENS BY
DUNCAN PHYFE, ETC., ETC.

TO BE SOLD BY HIS ORDER
FRIDAY, SATURDAY AFTERNOONS
APRIL ELEVENTH, TWELFTH
AT TWO-THIRTY O'CLOCK

THE ANDERSON GALLERIES

[MITCHELL KENNERLEY, President]
PARK AVENUE AND FIFTY-NINTH STREET, NEW YORK
1924

CONDITIONS OF SALE

1. All bids to be PER LOT as numbered in the catalogue.

2. The highest bidder to be the buyer; in all cases of disputed bids the lot shall be resold, but the Auctioneer will use his judgment as to the good faith of all claims and his decision shall be final. He also reserves the right to reject any fractional or nominal bid which in his judgment may delay or injuriously affect the sale.

3. Buyers to give their names and addresses and to make such cash payments on account as may be required, in default of which the lots purchased to be resold immediately.

4. The lots to be taken away at the buyer's expense and risk within twenty-four hours from the conclusion of the sale, and the remainder of the purchase money to be absolutely paid on or before delivery, in default of which The Anderson Galleries, Incorporated, will not be responsible if the lot or lots be lost, stolen, damaged, or destroyed, but they will be left at the sole risk of the purchaser, and subject to storage charges.

5. To prevent inaccuracy in delivery, and inconvenience in the settlement of purchases, no lot will be delivered during the sale.

6. All lots will be exposed for public exhibition in The Anderson Galleries before the date of sale, for examination by intending purchasers, and The Anderson Galleries, Incorporated, will not be responsible for the correctness of the description, authenticity, genuineness, or for any defect or fault in or concerning any lot, and makes no warranty whatever, but will sell each lot exactly as it is, WITHOUT RECOURSE. But upon receiving before the date of sale, expert opinion in writing that any lot is not as represented, The Anderson Galleries, Incorporated, will use every effort to furnish proof to the contrary, and in default of such proof the lot will be sold subject to the declaration of the aforesaid expert, he being liable to the owner or owners thereof for damage or injury occasioned by such declaration.

7. TERMS CASH. Upon failure to comply with the above conditions any sum deposited as part payment shall be forfeited, and all such lots as remain uncleared after twenty-four hours from the conclusion of the sale, will be resold by either private or public sale at such time as The Anderson Galleries, Incorporated, shall determine, without further notice, and if any deficiency arises from such resale it shall be made good by the defaulter at this sale together with all the expenses incurred thereby. This condition shall be without prejudice to the right of The Anderson Galleries, Incorporated, to enforce the contract with the buyer, without such re-sale.

8. BIDS. We make no charge for executing orders for our customers and use all bids competitively, buying at the lowest price permitted by other bids.

9. The Anderson Galleries, Incorporated, will afford every facility for the employment of carriers and packers by the purchasers, but will not be responsible for any damage arising from the acts of such carriers and packers.

Priced Copy of the Catalogue may be secured for fifty cents
for each session of the sale

THE ANDERSON GALLERIES
INCORPORATED
PARK AVENUE AND FIFTY-NINTH STREET, NEW YORK

TELEPHONE, PLAZA 9356   CATALOGUES ON REQUEST

SALES CONDUCTED BY MR. F. A. CHAPMAN AND MR. A. N. BADE

INTRODUCTION

WHILE I was at work putting the chairs, tables, chests of drawers, and highboys of my present fourth sale of American furniture in good condition, I had many friends and customers drop in at my workshop, which is open to everyone who wants to get an insight into my way of repairing and re-conditioning antique furniture.

There were also some dealers who came and said, "Say, Jake, be a good sport and let me have this piece. I will pay you a good price". However, according to my principle, I did not sell and that is why the present sale contains rare specimens such as No. 245, a 17th century maple day-bed, or No. 230, a Heppelwhite secretary-desk, and some other pieces which maybe will bring even a few dollars more at auction than I was offered before the sale.

I have often been asked another question: "How is it that you get a higher price for a piece of furniture, than an identical piece brings elsewhere?"

My son, who has just been through a course of merchandising, would say: "It is on account of the discriminating price-finding of the public." The customers have found out that a chest of drawers in perfect order at three hundred and fifty dollars is cheaper than the same chest of drawers "in the rough" at two hundred and fifty dollars plus a repair bill of one hundred and seventy-five dollars in addition to the "basic cost". Being only a cabinet maker, I say that my customers like to know where they stand; they want to get a genuine piece, and they want to know what it really will cost.

It is unnecessary to add, that I guarantee every piece of this sale to be genuine as described. I also shall be glad to make necessary minor repairs without charge for my New York customers.

JACOB MARGOLIS
Cabinet Maker

FIRST SESSION

NUMBERS 1-139

1 CURLY MAPLE AND MAHOGANY WALL MIRROR

EARLY AMERICAN, ABOUT 1800

Outer and inner moulding in light mahogany.

30

Size, 16 x 22 inches

2 PAIR OF EARLY AMERICAN BRASS ANDIRONS

25

Baluster columns. On bracket supports. (2)

3 SMALL QUEEN ANNE WALNUT WALL MIRROR

EARLY AMERICAN, ABOUT 1710

25 Arched top, moulded frame.   Size, 21 x 9¼ inches

4 PAIR OF EARLY AMERICAN BRASS ANDIRONS

Baluster pedestal with ball finials. Bracket feet. (2)

Height, 15 inches

20

5 CHIPPENDALE MAHOGANY WALL MIRROR

EARLY AMERICAN, ABOUT 1775

50 Top and bottom deeply scrolled and scalloped.

27

Size, 25½ x 13 inches

6 CURLY MAHOGANY WALL MIRROR

50

EARLY AMERICAN, ABOUT 1800

Rectangular frame with fine ogee moulding.

22

Size, 33½ x 22 inches

7 HEPPELWHITE INLAID MAHOGANY SHAVING MIRROR

50

EARLY AMERICAN, ABOUT 1790

The mirror in very graceful shield-shaped frame, with linear satin-

42 wood inlay around the edge. Shaped bracket supports.

8 MAPLE CANDLE STAND   EARLY AMERICAN, ABOUT 1770

35 Vase-shaped baluster support. Three outcurved legs ending in
snake-head feet.

9 CHILD'S HICKORY AND PINE ARMCHAIR

EARLY AMERICAN, ABOUT 1780

Five slats with bamboo turnings. High arms. Round raked legs with bamboo markings, connected by stretchers.

10 HICKORY ROCKING CHAIR EARLY AMERICAN, ABOUT 1800

Three-ladder back. Round posts. Rush seat. Round legs with double bracing.

11 CHIPPENDALE MAHOGANY WALL MIRROR

EARLY AMERICAN, ABOUT 1780

Arched and scrolled top, scalloped base.

Height, 32 inches; width, 16 inches

12 EARLY AMERICAN MAPLE COAT RACK

Turned column.

13 CHIPPENDALE MAHOGANY WALL MIRROR

EARLY AMERICAN, ABOUT 1780

Scrolled and scalloped top and base executed with the saw. Moulding next to the mirror in gilt.

Height, 30½ inches; width, 17 inches

14 OCTAGONAL MAPLE CANDLE STAND

EARLY AMERICAN, ABOUT 1780

Turned, urn-shaped support with square outcurved legs.

Size of top, 22 x 14 inches

15 PINE MINIATURE PANELLED CHEST

EARLY AMERICAN, ABOUT 1780

Drop lid, enclosing interior with one compartment. The front with sunken panel. Bracket base. Length, 30 inches

16 OVAL MAPLE TIP-TOP TABLE EARLY AMERICAN, ABOUT 1770

Urn-shaped baluster support. Three slender outcurved legs ending in spade feet. Size of top, 23½ x 14½ inches

17 PAIR OF HICKORY SIDE CHAIRS

EARLY AMERICAN, ABOUT 1800

Three-slat back. Turned posts. Rush seat. Round legs connected by braces. (2)

18 PINE AND MAPLE OCTAGONAL OCCASIONAL TABLE

EARLY AMERICAN, ABOUT 1740

Elongated octagonal top. Turned legs, connected by plain stretchers. Size of top, 30 x 19½ inches

85

19 CHIPPENDALE MAHOGANY WALL MIRROR

EARLY AMERICAN, ABOUT 1780

Scrolled and scalloped top and base. Inner frame moulded.

Height, 27 inches; width, 12 inches

30

20 PINE AND MAPLE OCCASIONAL TABLE

EARLY AMERICAN, ABOUT 1790

Rectangular top. Apron fitted with drawer. Square, slender tapering legs. Size of top, 18 x 15½ inches

45

21 HICKORY AND MAPLE SIDE CHAIR

EARLY AMERICAN, ABOUT 1765

Turned posts with low arched rail and baluster-shaped splat. Round legs with ring turning, connected in front by double turned stretchers, and plain stretchers at the sides.

35

22 EARLY AMERICAN MAHOGANY TIP-TOP TABLE

Scalloped quatrefoil top with urn-shaped baluster support and three outcurved legs. Size of top, 26 x 18 inches

55

23 CHERRY AND MAPLE CANDLE STAND

EARLY AMERICAN, ABOUT 1770

Oval top in cherrywood. Vase-shaped baluster support in maple. Three graceful outcurved legs ending in spade feet.

40

24 CHILD'S PAINTED ROCKING CHAIR

PENNSYLVANIA, ABOUT 1830

Wide top rail, painted with leaves. Scrolled arms. Sleigh front seat, with painted design. Unusual piece.

20

25 PINE AND MAPLE TAVERN TABLE

EARLY AMERICAN, ABOUT 1740

Pine top; turned legs, connected by plain stretchers. Ball feet.

Size of top, 29 x 20 inches

55

26 ROUND CANDLE STAND   EARLY AMERICAN, ABOUT 1760

Round top, supported by graceful turned baluster, with tripod base ending in snake-head feet. Oak, pine and maple.

40

27 HICKORY AND MAPLE WINDSOR ARMCHAIR

EARLY AMERICAN, ABOUT 1760

Seven-spindle back, the spindles piercing the roundabout arm rail which has turned supports at the ends. Saddle seat. Turned and raked legs with turned bracing.

28 MAHOGANY SEWING TABLE   EARLY AMERICAN, ABOUT 1800

Two drawers; turned legs, tapering slightly toward the feet.

29 HICKORY AND MAPLE SIDE CHAIR

EARLY AMERICAN, ABOUT 1730

Bow-shaped top rail with eared ends. Baluster-shaped splat. Rush seat with wooden corners. Legs very elaborately turned, the front legs connected by stretcher with vase and ring turning.

30 PINE AND MAPLE TABLE   EARLY AMERICAN, ABOUT 1710

Slender turned legs. Straight massive braces.

Size of top, 21 x 29½ inches

31 HICKORY "CARVER" CHAIR

A copy of the famous Early American type. High turned back posts and legs. Rush seat. (Reproduction)

32 CHILDS' HICKORY AND MAPLE LADDER-BACK ARM-CHAIR   EARLY AMERICAN, ABOUT 1770

Turned posts connected by two slightly arched slats. High arms. Turned legs connected by braces. Rush seat.

33 MAPLE SEWING TABLE   EARLY AMERICAN, ABOUT 1800

Two drawers. Turned legs tapering toward the feet.

34 PINE AND MAPLE TABLE   EARLY AMERICAN, ABOUT 1710

Apron fitted with one drawer. Double vase turned legs with plain square stretchers.   Size of top, 34 x 21 inches

35 HICKORY AND MAPLE WINDSOR COMB-BACK SIDE CHAIR   EARLY AMERICAN, ABOUT 1760

Six spindles. Unusually narrow seat. Turned and raked legs connected by bulbous stretchers.

36 MAPLE AND CURLY MAPLE OCCASIONAL TABLE

EARLY AMERICAN, ABOUT 1820

Apron with one drawer in fine curly maple. Turned baluster pedestal supported by four bracket feet scrolled at the top. Unusual piece.

37 PAIR OF HEPPELWHITE INLAID MAHOGANY KNIFE
BOXES   EARLY AMERICAN, ABOUT 1790
Edged with satinwood and ebony inlay. Serpentine fronts. With
original mountings. (2)

38 PINE AND MAPLE OCTAGONAL TAVERN TABLE
EARLY AMERICAN, ABOUT 1730
Elongated octagonal top with grooved edge. Turned and raked
legs, connected by plain stretchers. Size of top, 29½ x 21 inches

39 PINE AND MAPLE TAVERN TABLE
EARLY AMERICAN, ABOUT 1740
Apron with drawer; turned legs, connected by plain stretchers.
Size of top, 42 x 27½ inches

40 SMALL APPLEWOOD CHEST OF DRAWERS
EARLY AMERICAN, ABOUT 1780
Four drawers, increasing in depth towards the base, beaded at the
edge. Moulded base with bracket feet.
Height, 32 inches; length, 33 inches

41 MAPLE TAVERN TABLE   EARLY AMERICAN, ABOUT 1710
Apron fitted with one drawer. Turned legs connected by plain
square bracing.   Size of top, 34½ x 21½ inches

41A PENNSYLVANIA TAVERN TABLE
EARLY AMERICAN, ABOUT 1720
An unusual and rare piece. Pine and oak. The apron with
one drawer. Cabriole legs ending in unusual club and ball feet.
Height, 28 inches; size of top, 25 x 39 inches

42 PINE AND MAPLE TIP-TOP TABLE
EARLY AMERICAN, ABOUT 1775
Moulded dish top in pine. Turned pedestal support. Three legs
ending in snake-head feet.

43 MAPLE BUREAU   EARLY AMERICAN, ABOUT 1770
The four drawers set into beaded moulding. Slender bracket feet.
Height, 33 inches; length, 38 inches

5

44 RARE CHERRY AND MAPLE TIP-TOP TABLE

EARLY AMERICAN, ABOUT 1760

The top with finely moulded edge on birdcage support with small turned columns. The pedestal of turned maple with three well shaped legs ending in snake-head feet.   Diameter, 22 inches

[SEE ILLUSTRATION]

45 HICKORY AND MAPLE WINDSOR FAN-BACK ARMCHAIR

EARLY AMERICAN, ABOUT 1750

Fine and rare specimen. Fan-shaped back with nine turned spindles and two supporting spindles at the back resting on an extension of the seat. High scrolled arms. Finely turned legs connected by bulbous stretchers.

[SEE ILLUSTRATION]

46 HICKORY AND MAPLE WINDSOR FAN-BACK ARMCHAIR

EARLY AMERICAN, ABOUT 1750

Mate to the preceding.

[SEE ILLUSTRATION]

47 CURLY MAPLE BREAKFAST TABLE

EARLY AMERICAN, ABOUT 1790

Square tapering legs. Two square drop leaves.

Size of top, 40 x 41½ inches

48 APPLEWOOD SWELL-FRONT BUREAU

EARLY AMERICAN, ABOUT 1790

Four drawers in beaded frame. French bracket feet.

Height, 35 inches; length, 42 inches

49 MAPLE AND PINE TAVERN TABLE

EARLY AMERICAN, ABOUT 1710

Apron fitted with one drawer. Simply turned legs connected by plain square bracing.   Size of top, 32 x 21 inches

50 MAPLE BUREAU   EARLY AMERICAN, ABOUT 1780

Moulded top. Body with four drawers increasing in depth toward the base. Bracket feet.   Height, 39 inches; length, 37½ inches

HICKORY AND MAPLE
FAN-BACK ARMCHAIR
ABOUT 1750

[NUMBER 45]

CHERRY AND MAPLE
TIP-TOP TABLE
ABOUT 1760

[NUMBER 44]

HICKORY AND MAPLE
FAN-BACK ARMCHAIR
ABOUT 1750

[NUMBER 46]

51 HICKORY AND MAPLE LADDER-BACK ARMCHAIR

EARLY AMERICAN, ABOUT 1760

105

Turned posts with finely turned ball finials. Four-ladder back. High, slightly curved arms supported by the turned legs. Two front stretchers with vase and ring turnings. Plain double braces at side. Rush seat.

52 MAPLE SEWING TABLE

EARLY AMERICAN, ABOUT 1800

80

The square top with outcurved and scalloped corners. Two drawers. Round turned legs connected by scalloped stretcher.

Size of top, 19 inches square

53 HICKORY AND MAPLE ARMCHAIR

EARLY AMERICAN, ABOUT 1735

85

Four arched ladder splats. Round posts with turned finials. High arms with knuckled ends having turned supports. Round legs on ball feet. Front stretcher with massive ball and ring turning. Rush seat.

54 MAPLE CHEST OF DRAWERS EARLY AMERICAN, ABOUT 1790

145

The top with rounded corners. Four drawers; round pilasters at sides, ending in turned feet.

Height, 37½ inches; length, 40½ inches

55 MAHOGANY PEMBROKE TABLE

EARLY AMERICAN, ABOUT 1790

57 50

Apron with one drawer and old brass with three-feather emblem of the Prince of Wales. Slender, square tapering legs.

Size when open, 29½ x 34½ inches

56 HICKORY AND MAPLE COMB-BACK WINDSOR ARM-CHAIR

EARLY AMERICAN, ABOUT 1760

145

With nine spindles piercing the roundabout arm rail. Shaped ends with turned supports. Saddle seat. Turned and raked legs, connected by stretcher with bulbous turning.

57 HICKORY AND MAPLE WINDSOR ARMCHAIR

EARLY AMERICAN, ABOUT 1760

105

The fan-shaped back with six spindles which pierce the roundabout arms. Saddle seat; turned and raked legs with bulbous turned stretchers.

58 MAPLE BUREAU
EARLY AMERICAN, ABOUT 1780

Top with moulded edge. Four drawers. Bracket feet.

Height, 37 inches; length, 37 inches.

59 PINE AND MAPLE TAVERN TABLE
EARLY AMERICAN, ABOUT 1750

Pine top, turned legs connected by plain stretchers. Apron with drawer.

Size of top, 28½ x 22 inches.

60 MAPLE AND HICKORY DOUBLE-BACK WAGON SEAT
EARLY AMERICAN, ABOUT 1750

Round posts and double ladder back. Straight arms supported by turned posts. Double bracing at sides and front. Rush seat.

61 HEPPELWHITE MAHOGANY SEWING TABLE
EARLY AMERICAN, ABOUT 1790

Very graceful piece. The drop lid enclosing interior with numerous compartments. Drawer and sliding sewing bag below. Slender square legs with cross stretchers and shelf in centre. The drawers with ebony piping.

62 HICKORY AND MAPLE WINDSOR ARMCHAIR
EARLY AMERICAN, ABOUT 1765

Finely shaped comb-back with scrolled ends; eight spindles, pierced by the roundabout arm supports. Turned and raked legs connected by stretchers.

63 MAPLE FOUR-POSTER BED
EARLY AMERICAN, ABOUT 1800

Low head and foot board; the head board scrolled. Well turned posts. The posts on either side of the foot board fluted.

Width, 53 inches

64 MAPLE FALL-FRONT WRITING DESK
EARLY AMERICAN, ABOUT 1780

The fall front enclosing interior with eight large and small pigeon-holes, four drawers and central sunburst carved door enclosing locker. The lower body with four drawers and ogee bracket feet.

Height, 42 inches; length, 37 inches

65 INLAID HEPPELWHITE PEMBROKE TABLE

EARLY AMERICAN, ABOUT 1790

The top, which is oval when extended, is outlined by a narrow satinwood line. The apron similarly framed, and with edging in ebony with satinwood piping. Above the square, tapering legs, oval shell inlay, and below pendent husks. Fine specimen.

Size of top, 40½ x 33 inches

66 WALNUT QUEEN ANNE SIDE CHAIR

EARLY AMERICAN, ABOUT 1715

Open back with vase-shaped splat. Very wide seat. Cabriole legs ending in club feet. Elaborately turned stretcher.

67 LARGE WALNUT STRETCHER-TOP TABLE

EARLY AMERICAN, ABOUT 1730

Unusual type of tavern table, the apron with three drawers having original brasses; turned legs, connected by stretchers.

Size of top, 29 x 53½ inches

68 MAPLE LOW POSTER-BED EARLY AMERICAN, ABOUT 1810

Low head board with scrolled ends. The posts elaborately turned, the foot posts carved with acanthus leaves. Ball finials.

Width, 53 inches

69 WALNUT FALL-FRONT WRITING DESK

EARLY AMERICAN, ABOUT 1760

The fall front enclosing pigeonholes, small drawers and locker. Two pulls. Lower body with four wide drawers, increasing in depth towards the base. Old brasses. Unusual inlaid escutcheons and chamfered corners with line inlay. French feet.

Height, 41 inches; length, 39 inches

70 PAIR OF CHIPPENDALE MAHOGANY RECTANGULAR STOOLS EARLY AMERICAN, ABOUT 1770

Square legs, with reeded corners, connected by plain, square braces. Original upholstery, fastened with brass nails. (Legs restored) (2) Size of top, 23 x 16 inches

71 DUNCAN PHYFE MAHOGANY SERVING TABLE

EARLY AMERICAN, ABOUT 1810

Top with finely reeded edge. The apron fitted with two drawers framed by a mahogany frieze. Round legs with brass feet.

Size of top, 31 x 19 inches

72 MAHOGANY SHERATON BUREAU

EARLY AMERICAN, ABOUT 1800

90 — The top with rounded corners. Four drawers with beaded edge. Fluted pilasters at sides. Turned legs. Scalloped apron.

Height, 40 inches; length, 41½ inches

73 MAPLE FOUR-POSTER BED   EARLY AMERICAN, ABOUT 1800

Low head and foot board; the head board scrolled. Well turned *115 —* posts. The posts on either side of the foot board fluted.

Width, 53 inches

74 MAPLE FALL-FRONT WRITING DESK

EARLY AMERICAN, ABOUT 1780

165 — The fall front enclosing interior with seven drawers, six pigeon-holes, and bill drawers with unusual turned pilasters in front. Lower body with four drawers and bracket feet.

Height, 41 inches; length, 32 inches

75 MAHOGANY DROP-LEAF DINING ROOM TABLE

EARLY AMERICAN, ABOUT 1790

140 — Wide drop leaf. Plain apron. Square, slightly tapering legs.

Size when extended, 54 x 62 inches

76 VERY FINE HEPPELWHITE INLAID CURLY MAPLE AND MAHOGANY BUREAU   EARLY AMERICAN, ABOUT 1790

The edge of top framed by satinwood inlay. Four swell-front *150 —* drawers in curly maple, framed and panelled in mahogany inlay and edged in satinwood. Deeply scalloped apron with rectangle of maple in the centre. Slender French bracket feet. Very fine piece.   Height, 38 inches; length, 41½ inches

77 MAPLE AND PINE KITCHEN DRESSER

95 — EARLY AMERICAN, ABOUT 1800

The upper body with shelf and two drawers connected with the lower body by scrolled bracket-end supports. Lower body enclosed by door with sunken panel. Bracket feet.

Height, 49½ inches; length, 29 inches

78 MAPLE CHEST ON FRAME   EARLY AMERICAN, ABOUT 1720

170 — Five drawers. Finely scalloped apron. Curved legs ending in club feet.   Height, 57½ inches; length, 38½ inches

10

RARE HICKORY AND MAPLE SPANISH FOOT ARMCHAIRS
EARLY AMERICAN, ABOUT 1720

[NUMBER 84]   [NUMBER 85]

79 CHERRYWOOD CORNER CABINET

EARLY AMERICAN, ABOUT 1800

Moulded cornice. Four shelves enclosed by panel glass door; three arches at the top. Elaborately turned pilasters on the corners. Lower body with long drawer above two doors.

Height, 85 inches; length about 42 inches

80 WALNUT FALL-FRONT WRITING DESK

EARLY AMERICAN, ABOUT 1775

The fall front enclosing pigeonholes, small drawers and locker. The small drawers of unusually fine crotch walnut. Lower body with four drawers, increasing in depth towards the base. Ogee bracket feet. Height, 45 inches; length, 40 inches

81 PINE CUPBOARD   EARLY AMERICAN, ABOUT 1780

Deeply moulded top with dentelled frieze below. Two lattice glass doors enclosing interior with three shelves. Lower body with cupboard enclosed by two panelled doors. Bracket feet.

Height, 79 inches; length, 40 inches

82 MAPLE GATELEG TABLE

EARLY AMERICAN, LATE 17TH CENTURY

The top round when extended. Well turned legs connected by well turned braces. Apron with one drawer.

Diameter, 42 inches

83 MAPLE CHIPPENDALE FALL-FRONT WRITING DESK

EARLY AMERICAN, ABOUT 1780

Fall front opens to disclose writing interior with a row of eight pigeonholes at the top, above two rows of seven drawers. The base with four long drawers with wide moulding below. Very finely carved ball and claw feet. Unusually fine piece.

Height, 41 inches; length, 41 inches

84 RARE HICKORY AND MAPLE ARMCHAIR

EARLY AMERICAN, ABOUT 1710-20

Arched top rail with eared ends. Finely shaped violin splat. High arms scrolled at the ends and with turned supports. Finely turned legs ending in Spanish feet. Large front stretcher with vase and ring turning in centre. Double braces at sides. Rush seat.

[SEE ILLUSTRATION]

11

85 RARE HICKORY AND MAPLE SPANISH FOOT ARM-CHAIR
EARLY AMERICAN, ABOUT 1720

Finely shaped top rail. Slender baluster splat. Round arms beautifully scrolled at the ends. Elaborately turned legs ending in Spanish feet and connected by stretcher with double ball and ring turning. Double bracings at side.

[SEE ILLUSTRATION]

150

86 MAPLE HIGH CHEST OF DRAWERS
EARLY AMERICAN, ABOUT 1750

Moulded top. Six long drawers increasing in depth toward the base, which is finely moulded and has short cabriole legs ending in club feet. Height, 51 inches; length, 38 inches

260

87 CHILD'S RARE HEPPELWHITE MAHOGANY AND CURLY MAPLE BUREAU
EARLY AMERICAN, ABOUT 1790

Very unusual piece. The top with band of narrow inlay around the edge. Four long drawers increasing in depth toward the base. The centre panel in curly maple; the border in mahogany. French bracket feet and curved apron, with inlay just above.
Height, 24½ inches; length, 21 inches

80

88 MAPLE BUREAU
EARLY AMERICAN, ABOUT 1775

Top with moulded edge. Four long drawers. Moulded base with bracket feet. Height, 37 inches; length, 39 inches

125

89 CURLY MAPLE DESK
EARLY AMERICAN, ABOUT 1780

Very beautiful grained wood. The fall front enclosing interior with six pigeonholes and numerous small drawers, some of them scalloped. Compartment with three drawers in centre enclosed by door. Lower body with four drawers increasing in depth toward the base. Bracket feet. Height, 41 inches; length, 36 inches

160

90 PAIR OF HEPPELWHITE INLAID MAHOGANY SIDE CHAIRS
EARLY AMERICAN, ABOUT 1790

Open shield-shaped back, with curved top, with three slender slats, inlaid with oval fan medallions in satinwood and with half fan, at the base of the shield above the back. Upholstered seat. Square tapering legs, with narrow satinwood inlay. (2)

[SEE ILLUSTRATION]

220

HEPPELWHITE INLAID FOLD-TOP CARD TABLE AND PAIR OF
HEPPELWHITE INLAID MAHOGANY SIDE CHAIRS
EARLY AMERICAN, ABOUT 1790

[NUMBER 90]   [NUMBER 91]   [NUMBER 90]

91 HEPPELWHITE INLAID FOLD-TOP CARD TABLE

EARLY AMERICAN, ABOUT 1790

The top with cut and rounded corners, and with herringbone inlay around the end. The deep apron shaped at the corners and with linear inlay in satinwood, forming a framed oval. The lower edge of the apron with very unusual frieze in satinwood, etc. Graceful, square tapering legs, inlaid with lines.

Size of top, extended, 35 x 35 inches

[SEE ILLUSTRATION]

92 FINE WALNUT FALL-FRONT WRITING DESK

EARLY AMERICAN, ABOUT 1780

Very fine serpentine front interior with central locker with movable drawers and pigeonholes concealing secret drawers, bill files with turned pilasters in front and numerous finely finished serpentine front drawers and pigeonholes. The lower body with row of three small drawers across the top; three long drawers below. Ogee bracket feet. Height, 42½ inches; length, 37 inches

93 CURLY MAPLE HIGHBOY   EARLY AMERICAN, ABOUT 1760

Moulded cornice. The upper body with row of three small drawers, and below, a row of two small drawers, followed by three wide drawers. The upper part fits into the moulded base, the apron of which has a row of two small drawers and is finished with elaborate scalloping. Cabriole legs with pad feet.

Height, 62½ inches; length, 37½ inches

94 NEW ENGLAND PINE DRESSER

EARLY AMERICAN, ABOUT 1730

F. moulded top; three shelves below with grooves for plate. The ends with finely scalloped outline; lower body with grooved plate rail at back. Two panel doors inclosing a cupboard.

Height, 81 inches; length, 57 inches

95 SET OF FOUR MAHOGANY SIDE CHAIRS

EARLY AMERICAN, ABOUT 1820

Probably by Duncan Phyfe. Conforming and scroll back with panelled top rail and crossbar elaborately carved with flowers and leaves. Plain outcurved legs in one piece with the frame. Leather slip seats. (4)

96 MAPLE DROP-LEAF TABLE EARLY AMERICAN, ABOUT 1715-20
The drop leaves rounded. Apron finely scalloped at each end.
Round tapering legs ending in small club feet.

Diameter, when extended, 35 inches

97 MAPLE LOWBOY EARLY AMERICAN, ABOUT 1730
Plain pine top; body with one long drawer above a row of three
square drawers. Elaborate scalloped apron with two turned
urn-shaped drop motifs. Cabriole legs ending in club feet.

Height, 31 inches; length, 33 inches

[SEE ILLUSTRATION]

98 QUEEN ANNE MAPLE CHEST OF DRAWERS ON BASE
EARLY AMERICAN, ABOUT 1720

Moulded cornice. Five drawers increasing in width toward the
base. Finely scalloped apron. Round turned legs with round
feet. Fine early piece. Height, 58 inches; length, 39 inches

99 QUEEN ANNE CHERRYWOOD HIGHBOY
EARLY AMERICAN, ABOUT 1720

The upper body with central square drawer carved with sunburst
flanked on each side by unusual arrangement of two small drawers. Four long drawers below. Upper body sets into frame with
apron having two drawers. Cabriole legs with club feet.

Height, 66 inches; length, 37 inches

100 CHIPPENDALE MAHOGANY WING BACK ARMCHAIR
EARLY AMERICAN, ABOUT 1770

Chinese Chippendale square front legs, with relief carving of
leaves and fretwork. Plain square braces. Upholstered in blue
denim.

101 PINE AND MAPLE TAVERN TABLE
EARLY AMERICAN, ABOUT 1730

Plain rectangular pine top. Maple base. Four round, slightly
raked legs connected by plain square stretchers. Fine specimen.

Size of top, 23½ x 15½ inches

[SEE ILLUSTRATION]

EARLY AMERICAN MAPLE LOWBOY—ABOUT 1730

[NUMBER 97]

102 PINE AND MAPLE TAVERN TABLE

EARLY AMERICAN, ABOUT 1730

The oval top in finely grained pine. Round turned legs, slightly raked and connected by plain stretcher. Apron fitted with one drawer. Size of top, 28 x 20 inches

[SEE ILLUSTRATION]

103 HEPPELWHITE INLAID MAHOGANY TAMBOUR-FRONT BUREAU DESK EARLY AMERICAN, ABOUT 1790

Tambour front enclosing three pigeonholes and the two drawers on each side. Centre compartment enclosed by fluted swell-front door. Lower body with swell front and four long drawers. The base with banded inlay of ebony and satinwood. French bracket feet. Height, 43 inches; length, 42 inches

104 DUNCAN PHYFE MAHOGANY CARD TABLE

EARLY AMERICAN, ABOUT 1810

The top with chamfered corners and hinged leaf. Bracket-end supports, each with two columns with spiral turnings; brass lion-claw casings on feet. The end brackets connected by spirally turned stretcher. Swinging top with secret drawer.

Size of top, 36 x 35 inches

105 MAPLE DROP-LEAF TABLE EARLY AMERICAN, ABOUT 1740

Plain top of rectangular lines with short drop leaves. Square legs connected by a long stretcher and with square bracing at the end. Size of top, extended, 65 x 35 inches

106 CURLY MAPLE HIGHBOY EARLY AMERICAN, ABOUT 1720

Moulded top. Upper body with five drawers increasing in depth toward the lower body. The lower body with moulded top into which the upper section sets. Long narrow drawer across the top, and below a small drawer flanked by two square drawers. Deeply scalloped apron with turned drops. Cabriole legs with club feet. Height, 68 inches; length, 37 inches

107 FINE HEPPELWHITE INLAID MAHOGANY THREE-PIECE DINING ROOM TABLE EARLY AMERICAN, ABOUT 1790

With narrow satinwood inlay and ebony frieze on the apron. Rectangular panels of satinwood above each of the legs, which are decorated below with pendent floral leaves. The end table with swell front. The centre table with unusual drop leaves supported by swinging leg, and perfectly finished to be used as a separate table.

Length when extended, 13 feet 2 inches; width, 4 feet 4 inches

15

108 PAIR OF CARVED BEECH FLEMISH-STYLE SIDE CHAIRS EARLY AMERICAN, 1690-1700

110 —

Scroll top rail with leaf ornament in centre. Elaborately turned posts with urn finial. Turned front legs connected by front stretcher with scroll carving. Back and seat upholstered in leather. (2)

109 CHIPPENDALE MAHOGANY WING-BACK ARMCHAIR EARLY AMERICAN, ABOUT 1775

195 —

Square legs, reeded on the corners, and connected by plain square braces. Upholstered in cretonne with Chinoiserie pattern.

110 SET OF FIVE CURLY MAPLE CHIPPENDALE · SIDE CHAIRS EARLY AMERICAN, ABOUT 1780

625 —

Very finely grained wood. Gracefully curved bow-shaped top rail and wide violin-shaped splat. The square legs connected by plain braces. Very fine set. Slip seats. (5)

111 MAPLE HIGHBOY EARLY AMERICAN, ABOUT 1730

300 —

Unusually small size. Moulded cornice above a row of three small drawers, the centre one of which has fine sunburst carving. Three large drawers below. The lower body with a long drawer above a row of three square drawers, the centre one having sunburst carving. Deeply scalloped apron. Cabriole legs and club feet. Height, 66 inches; length, 39 inches

112 FINE HEPPELWHITE INLAID MAHOGANY THREE-PIECE DINING ROOM TABLE EARLY AMERICAN, ABOUT 1790

210 —

The two end sections semi-circular; the centre table rectangular, perfectly finished across the front, making it suitable for use as a separate console table. The aprons inlaid with satinwood and bordered by a mahogany frieze. Above the square tapering legs satinwood inlay simulating fluting. The end pieces with swell front. Length, 72 inches; greatest width, 47½ inches

113 WALNUT LOWBOY EARLY AMERICAN, ABOUT 1730

875 —

Top with moulded edge, the apron with row of three drawers and elaborate scalloping. Very unusual cabriole legs, with fluted Spanish feet. Original handles. Very unusual piece on account of the design of the legs. Scalloped sides.

Height, 28 inches; size of top, 34 x 20 inches

[SEE ILLUSTRATION]

EARLY AMERICAN PINE AND MAPLE TAVERN TABLES—ABOUT 1730

[NUMBER 101]   [NUMBER 102]

114 MAPLE HIGHBOY EARLY AMERICAN, ABOUT 1740

Plain moulded top; the upper body with row of two small drawers above four long drawers, sets into the moulding of the lower body, which has a row of two drawers across the top. The second row of drawers has in the centre a rectangular drawer with fine sunburst carving flanked by two small drawers. Two turned drop ornaments. Cabriole legs with club feet.

Height, 71 inches; length, 38 inches

115 FINE HEPPELWHITE INLAID MAHOGANY PEMBROKE TABLE EARLY AMERICAN, ABOUT 1790

The top, which is oval when extended, is outlined by a narrow satinwood and ebony inlay. The apron with one drawer inlaid similarly to the top, and finished around the edge with ebony and satinwood. Square tapering legs inlaid at the top with three satinwood lines and below with pendent leaves. Fine specimen.

Size of top, 36½ x 32 inches

116 FINE MAPLE BUREAU EARLY AMERICAN, ABOUT 1780

The four drawers set into beaded moulding. Finely moulded base. Bracket feet. Height, 35 inches; length, 40 inches

117 FINE DUNCAN PHYFE MAHOGANY SEWING TABLE
EARLY AMERICAN, ABOUT 1810

The upper body with turned pilasters at the corners and two drawers with beaded edge. The pedestal support with elaborate acanthus carving. Four outcurved legs with the original brass lion-claw casings. Original brass knobs.

118 CHIPPENDALE MAHOGANY MIRROR
EARLY AMERICAN, ABOUT 1770

Scrolled top with openwork gilt rocaille leaf ornament and gilt sprays of leaves. The inner moulding scalloped and gilt. A charming piece. Size, 35 x 19½ inches

119 MAPLE AND HICKORY DOUBLE-BACK WAGON SEAT
EARLY AMERICAN, ABOUT 1750

Massive round posts with three-ladder back. Round straight arms supported by the high legs. Double bracing at front and sides. Rush seat.

17

120 HICKORY AND MAPLE LADDER-BACK ARMCHAIR

EARLY AMERICAN, ABOUT 1760

Five very finely arched vertical ladder slats set between the turned posts which end in ball finials. Straight arms with turned supports. Large front stretcher with vase and ring turnings connecting the round legs. Rush seat.

121 MAPLE AND BIRCH CHEST OF DRAWERS

EARLY AMERICAN, ABOUT 1780

Top shelf with moulded edge. Four drawers, increasing in depth towards the base. Moulded base, with bracket feet.

Height, 32½ inches; length, 39 inches

122 HICKORY AND MAPLE LADDER ROCKING CHAIR

EARLY AMERICAN, ABOUT 1735

Four arched ladder back connecting the turned posts which end in ball finials. The arms supported by unusual turned braces running down to the side stretcher. Plain round legs and side braces, double-vase turned front stretcher. Rush seat.

123 OVAL MAPLE TIP-TOP TABLE EARLY AMERICAN, ABOUT 1775

Very graceful piece with finely shaped oval top, supported by slender urn-shaped baluster; three outcurved legs in curly maple ending in snake-head feet. Size of top, 23½ x 16 inches

124 HICKORY AND MAPLE SIDE CHAIR

EARLY AMERICAN, ABOUT 1720

Shaped top rail supported by baluster splat. Rush seat with wooden corner pieces. Elaborately turned front legs, ending in well-carved Spanish feet and connected by stretcher with double vase and ring turning.

124A CHIPPENDALE POLE SCREEN

EARLY AMERICAN, ABOUT 1760

Needlework screen. Tripod base of outcurved legs, terminating in Dutch feet.

Height, 59 inches; size of screen, 17½ x 14 inches

125 PINE AND MAPLE TAVERN TABLE WITH SLIDING SHELF
EARLY AMERICAN, ABOUT 1715

Apron with long drawer and sliding shelf below. Elaborately turned legs connected by plain stretchers.

Size of top, 40 x 24 inches

EARLY AMERICAN WALNUT LOWBOY—ABOUT 1730

[NUMBER 113]

126 EARLY AMERICAN IRON CANDLE STAND
Horizontal branch with two candle sconces. Tripod base.

45—

127 MAPLE CANDLE STAND   EARLY AMERICAN, ABOUT 1770
Square top with chamfered and knuckled corners. Slender vase-
shaped support with three graceful outcurved legs.

45—

**128 HICKORY AND MAPLE COMB-BACK WINDSOR SIDE
CHAIR**   EARLY AMERICAN, ABOUT 1760
Six spindles flanked by turned posts. Saddle seat. Turned and
raked legs connected by bulbous stretchers. With unusual
maker's name, "B. Green", carved under the seat.

50

129 OVAL MAHOGANY TIP-TOP TABLE
EARLY AMERICAN, ABOUT 1780
The gracefully shaded top with urn-shaped baluster support and
three outcurved legs ending in spade feet.

Size of top, 24 x 15 inches

55

130 PINE AND MAPLE TAVERN TABLE
EARLY AMERICAN, ABOUT 1740
Pine top; turned legs connected by plain stretchers.

Size of top, 23½ x 33 inches

75

131 MAHOGANY CHIPPENDALE WALL MIRROR
EARLY AMERICAN, ABOUT 1780
The top and base with elaborate scrolls and scallops executed
with the saw.   Height, 28½ inches; width, 14 inches

30

132 APPLEWOOD TIP-TOP TABLE   EARLY AMERICAN, ABOUT 1800
Shaped and slightly scalloped top. Urn-shaped pedestal support
with three outcurved bracket legs.

55

133 HEPPELWHITE INLAID MAHOGANY CELLARET
EARLY AMERICAN, ABOUT 1790
Square shape, the entire body outlined by satinwood linear inlay. Short square legs, with inlaid band above the feet. Interior
with compartments for twelve bottles. Size of top, 16 x 12 inches

55

134 PINE AND MAPLE OCCASIONAL TABLE
EARLY AMERICAN, ABOUT 1740
Plain top, supported by four plain turned maple legs, connected
by stretchers.   Size of top, 22 x 17½ inches

60

135 INLAID MAHOGANY HEPPELWHITE SHAVING STAND

ENGLISH, ABOUT 1790

Shield-shaped swinging mirror outlined by satinwood inlay. Swell-front base with three drawers and bracket feet.

Height, 25 inches; length, 15½ inches

136 POPLAR AND APPLEWOOD OCCASIONAL TABLE

EARLY AMERICAN, ABOUT 1790

Top in applewood; the apron in poplar, forming a charming contrast. Square, tapering applewood legs.

Size of top, 15¼ inches square

137 UNUSUAL CURLY MAPLE DOLLS' SIDEBOARD

EARLY AMERICAN, ABOUT 1800

Apron with two small drawers. Bottom with two panel doors enclosing interior with finely scalloped sliding shelves and scalloped compartment. Turned pilasters at sides. Original Sandwich glass knobs. Height, 13 inches; length, 14½ inches

138 CHERRY AND MAPLE CANDLE STAND

EARLY AMERICAN, ABOUT 1770

Square top of wild cherrywood. Turned vase-shaped support in maple, with three outcurved legs.

139 PINE AND MAPLE OCCASIONAL TABLE

EARLY AMERICAN, ABOUT 1790

Rectangular pine top. Maple base. Apron fitted with one drawer. Square, slightly tapering legs. Size of top, 19½ x 15 inches

SECOND SESSION

NUMBERS 140-266

140 EARLY AMERICAN FOOTSTOOL   ABOUT 1820
Turned legs; upholstered seat.

141 HEPPELWHITE MAHOGANY SHAVING MIRROR
EARLY AMERICAN, ABOUT 1800
The square mirror supported between turned posts with pointed finials. Body with one drawer, set in frame with satinwood inlay.

142 PAIR OF EARLY AMERICAN BRASS ANDIRONS
Baluster-shaped columns on bracket feet. (2)   Height, 17 inches

143 SMALL MAHOGANY BRACKET MIRROR
EARLY AMERICAN, ABOUT 1790
The oval mirror between bracket supports.

144 MAPLE OCCASIONAL TABLE   EARLY AMERICAN, ABOUT 1790
The square top with rounded corners; apron with one drawer. Square tapering legs.

145 MINIATURE WALNUT CHEST   EARLY AMERICAN, ABOUT 1770
Drop lid enclosing interior with covered compartments. Moulded base with bracket feet.   Height, 15 inches; length, 30 inches

146 MAPLE OCCASIONAL TABLE   EARLY AMERICAN, ABOUT 1790
Apron with one drawer; square tapering legs; the wood of slightly curly grain.

147 CHIPPENDALE MAHOGANY WALL MIRROR
EARLY AMERICAN, ABOUT 1775
The deeply scrolled and scalloped top with gilt triple wheat ornaments in centre. Gilt inner moulding.   Size, 27 x 15 inches

148 MAPLE OCCASIONAL TABLE   EARLY AMERICAN, ABOUT 1790
Rectangular top. Apron fitted with one drawer. Square, slightly tapering legs.   Size of top, 18 x 14½ inches

149 HICKORY AND MAPLE WINDSOR ARMCHAIR

EARLY AMERICAN, ABOUT 1765

With seven spindles piercing the roundabout arm rail, which is shaped at the end and has a turned support. Saddle seat. Turned and raked legs with turned stretcher.

150 HEPPELWHITE INLAID MAHOGANY SHAVING STAND

EARLY AMERICAN, ABOUT 1790

Unusually small size. The rectangular mirror supported between elaborate turned posts. Swell front with one drawer, the frame edged in satinwood.

151 MAHOGANY SHERATON SHAVING MIRROR

EARLY AMERICAN, ABOUT 1800

The rectangular mirror supported between turned posts. The body with two long and two short drawers.   Length, 32 inches

152 PAIR OF HICKORY SIDE CHAIRS

EARLY AMERICAN, ABOUT 1800

Three-slat back. Turned post. Rush seat. Round legs connected by braces. (2)

153 CHIPPENDALE MAHOGANY WALL MIRROR

EARLY AMERICAN, ABOUT 1780

The arched and scalloped top with carved and gilt Chippendale phœnix in the centre. Scalloped bottom.

Height, 35 inches; width, 14 inches

154 CHERRY TIP-TOP TABLE   EARLY AMERICAN, ABOUT 1790

The square top with raised and beaded edge. Vase-shaped pedestal support with three outcurved legs.

155 UNUSUALLY EARLY AMERICAN PAINTED IRON SHAVING MIRROR

The oval mirror supported between two figures of ladies in full skirts standing on two branches of wheat. The American flag with 16 stars, Bunker Hill monument and the American shield below. Base with ships and flowers.   Height, 24 inches

156 MAHOGANY SEWING TABLE   EARLY AMERICAN, ABOUT 1790

Top with grooved edge. Body with two narrow drawers. Square, slightly tapering legs.

157 HICKORY AND MAPLE WINDSOR COMB-BACK SIDE
CHAIR   EARLY AMERICAN, ABOUT 1760
Seven turned spindles flanked on each side by delicately turned
posts. Saddle seat. Turned and raked legs connected by bulbous
turned stretcher.

158 SQUARE MAPLE OCCASIONAL TABLE
                                EARLY AMERICAN, ABOUT 1775
Top with beaded edge. Vase-shaped baluster support. Three
outcurved legs ending in snake-head feet.
Size of top, 17½ inches square

159 CHIPPENDALE MAHOGANY WALL MIRROR
                                EARLY AMERICAN, ABOUT 1780
Top and base elaborately scrolled and scalloped. The mirror in
moulded frame.   Height, 33½ inches; width, 17 inches

160 OVAL MAHOGANY AND MAPLE TIP-TOP TABLE
                                EARLY AMERICAN, ABOUT 1780
Ringed, urn-shaped baluster support. Three outcurved legs.
Top of finely grained mahogany.   Size of top, 21 x 15 inches

161 HICKORY AND MAPLE WINDSOR ARMCHAIR
                                EARLY AMERICAN, ABOUT 1750
Arched back with roundabout arm rail pierced by six spindles
and with two shorter spindles below. Finely knuckled arms supported by turned posts. Saddle seat. Turned legs connected by
stretchers with bulbous turnings.

162 MAHOGANY SEWING TABLE   EARLY AMERICAN, ABOUT 1790
Two drawers; round turned legs.

163 CHIPPENDALE INLAID MAHOGANY WALL MIRROR
                                EARLY AMERICAN, ABOUT 1780
Top and base with scrolls and scallops executed with the saw.
The mirror framed by gilt moulding and outer band of inlaid
mahogany.   Height, 31 inches; width, 13 inches

164 PINE AND MAPLE OCCASIONAL TABLE
                                EARLY AMERICAN, ABOUT 1740
Tavern type. Pine top. Turned legs, connected by stretchers.
Height, 22½ inches; size of top, 26 x 19½ inches

165 INLAID MAHOGANY WALL MIRROR

EARLY AMERICAN, ABOUT 1780

With part of New York label of manufacturer at 161 Fulton Street. The top and bottom with elaborate scrollwork executed with saw. The inner moulding with narrow satinwood inlay.

Height, 29 inches; width, 16 inches

166 SQUARE MAPLE BEDSIDE TABLE

EARLY AMERICAN, ABOUT 1790

Apron with one drawer having overlapping moulding. Square, slightly tapering legs.

Size of top, 16 inches square

167 HICKORY AND MAPLE LADDER-BACK SIDE CHAIR

EARLY AMERICAN, ABOUT 1750

Five arched ladder slats. Round posts with ball finials. Rush seat. Simply turned front legs connected by stretcher with double ball turning.

168 MAPLE SEWING TABLE EARLY AMERICAN, ABOUT 1800

Two drawers. Turned legs. Finely grained wood.

169 CHERRY CANDLE STAND EARLY AMERICAN, ABOUT 1790

Rectangular top with chamfered corners. Vase-shaped turned support with three slender legs.

170 PAIR OF HICKORY SIDE CHAIRS

EARLY AMERICAN, ABOUT 1800

Three-slat back. Turned posts. Rush seat. Round legs connected by braces. (2)

171 PAIR OF HICKORY SIDE CHAIRS

EARLY AMERICAN, ABOUT 1800

Three-slat back. Turned posts. Rush seat. Round legs connected by braces. (2)

172 PAIR OF HICKORY SIDE CHAIRS

EARLY AMERICAN, ABOUT 1800

Three-slat back. Turned posts. Rush seat. Round legs connected by braces. (2)

173 PAIR OF HICKORY SIDE CHAIRS

EARLY AMERICAN, ABOUT 1800

Three-slat back. Turned posts. Rush seat. Round legs connected by braces. (2)

174 OVAL APPLEWOOD CANDLE STAND

EARLY AMERICAN, ABOUT 1770

Turned vase-shaped support with three legs having snake-head feet.

175 UNUSUAL CURLY MAPLE SIDE CHAIR

EARLY AMERICAN, ABOUT 1810

Spirally carved top rail with pierced horizontal splat below. Wide rounded seat. Rounded legs decorated with a deep band of fluting. Cane seat.

176 MAPLE CANDLE STAND   EARLY AMERICAN, ABOUT 1770

Square top with scalloped corners. Turned baluster support with ringed urn at base and three slender outcurved legs ending in spade feet.

177 PINE AND CURLY MAPLE TAVERN TABLE

EARLY AMERICAN, ABOUT 1740

Double turned legs, connected by plain stretchers.

Size of top, 33 x 25½ inches

178 ROUND MAPLE CANDLE STAND

EARLY AMERICAN, ABOUT 1770

Turned baluster, vase-shaped support with three legs ending in snake-head feet.

179 PINE MIRROR WITH PAINTED GLASS INSETS

EARLY AMERICAN, ABOUT 1720

Arched and moulded top with painted glass panel having flower-basket decoration. The moulded sides with inset strip of painted glass. (Slightly damaged)   Size, 16 x 11 inches

180 HICKORY AND MAPLE LADDER-BACK ARMCHAIR

EARLY AMERICAN, ABOUT 1740

Four arched ladder slats; turned posts ending in ball finials. Low arms with turned supports. Round legs connected by stretcher with finely turned vase and ring in front.

181 MAPLE AND TAVERN TABLE EARLY AMERICAN, ABOUT **1710**
Pine top, maple base. Apron with large drawer. Legs with vase
and ring turning connected by plain massive stretchers.

Size of top, 40 x 24 inches

**182 PAIR OF AMERICAN EMPIRE TYPE MAPLE SIDE
CHAIRS** EARLY AMERICAN, ABOUT **1820**
Slightly scrolled top rail. Wide back rail below. Cane seats.
Turned legs with inturned brace in front. (2)

**183 PAIR OF AMERICAN EMPIRE TYPE MAPLE SIDE
CHAIRS** EARLY AMERICAN, ABOUT **1820**
Matching the preceding. (2)

**184 PAIR OF AMERICAN EMPIRE TYPE MAPLE SIDE
CHAIRS** EARLY AMERICAN; ABOUT **1820**
Matching the preceding. (2)

185 CURLY MAPLE BREAKFAST TABLE
EARLY AMERICAN, ABOUT **1790**
Square tapering legs. Two square drop leaves.

Size of top, 42 x 37 inches

186 PINE AND MAPLE TAVERN TABLE
EARLY AMERICAN, ABOUT **1710**
Pine top. Apron with drawer and moulded edge. Gracefully
turned legs connected by plain heavy stretchers.

Size of top, 41 x 23½ inches

**187 HICKORY AND MAPLE FAN-BACK WINDSOR ARM-
CHAIR** EARLY AMERICAN, ABOUT **1760**
Seven slender slats. Turned, slightly raked legs connected by
stretchers with bulbous turning.

188 PINE AND MAPLE TAVERN TABLE
EARLY AMERICAN, ABOUT **1730**
Plain top, supported by legs with double-vase turning and con-
nected by plain stretchers. Apron with drawer.

Top, 35 x 22 inches

189 HICKORY AND MAPLE WINDSOR ARMCHAIR

EARLY AMERICAN, ABOUT 1760

The fan back with seven spindles which pierce the roundabout arm supports. The arms with finely knuckled ends supported by turned posts. Turned and raked legs with bulbous stretchers.

190 CURLY MAPLE DROP-LEAF TABLE

EARLY AMERICAN, ABOUT 1800

Square leaves. Large drawer at one end. Legs with ring turnings.

Size of top, 47 x 45 inches

191 SMALL OAK TABLE
EARLY AMERICAN, ABOUT 1715

Unusual specimen, the top and drawer in pine. Turned legs connected by plain bracing. Size of top, 24 x 22 inches

192 CURLY MAPLE BUREAU
EARLY AMERICAN, ABOUT 1780

Three drawers of very finely grained curly maple. Graceful bracket feet. Height, 35 inches; length, 38½ inches

193 MAPLE TAVERN TABLE
EARLY AMERICAN, ABOUT 1740

Rectangular top, with unusual semi-circular scallopings in the four corners, inspired by the folding card table type. Turned legs, connected by stretchers. Size of top, 21½ x 32 inches

194 MAPLE CHEST OF DRAWERS
EARLY AMERICAN, ABOUT 1780

Top with moulded edge. Four drawers increasing in depth towards the base, which is moulded and has square bracket feet.

Height, 32 inches; length, 37½ inches

195 HICKORY AND MAPLE SIDE CHAIR

EARLY AMERICAN, ABOUT 1720

Very finely proportioned back with bow-shaped top and slender baluster splat. Rush seat with wooden corner pieces. Elaborately turned front legs ending in Spanish feet. Front stretcher with vase and ring turning.

196 APPLEWOOD FALL-FRONT WRITING DESK

EARLY AMERICAN, ABOUT 1780

The fall front enclosing interior with eight small pigeonholes, six drawers and central locker with drawers. Base with four drawers and bracket feet. Height, 43 inches; length, 39½ inches

197 MAPLE FOUR-POSTER BED   EARLY AMERICAN, ABOUT 1800

Low head and foot board; the head board scrolled. Well turned posts. The posts on either side of the foot board fluted.

Width, 53 inches

180

198 CURLY MAPLE CHEST   EARLY AMERICAN, ABOUT 1775

Drop lid enclosing interior with covered compartment to the right with two secret drawers. Base with one long drawer. Bracket feet.

Height, 32½ inches; length, 50 inches

165

199 HICKORY AND MAPLE ROCKING ARMCHAIR

EARLY AMERICAN, ABOUT 1760

Turned posts with urn finials. Baluster-shaped splat. Straight arms scrolled at the ends and supported by the high turned legs. Rush seat. Interesting piece.

60

200 MAPLE BUREAU   EARLY AMERICAN, ABOUT 1770

Four drawers gradually increasing in depth toward the base and set in overlapped moulding. Bracket feet.

Height, 35 inches; length, 38 inches

120

201 MAPLE DROP-LEAF TABLE   EARLY AMERICAN, 1715-20

Round drop leaves. The apron scalloped at each end. Round legs ringed at the top and with flat round feet.

Size of top, extended, 48 x 50 inches

202 SHERATON MAPLE BUREAU   EARLY AMERICAN, ABOUT 1790

Four drawers with beaded edge. The wood with slightly curly grain. Fluted corners. Turned legs.

Height, 38½ inches; length, 39 inches

110

203 PINE AND MAPLE TABLE   EARLY AMERICAN, ABOUT 1710

Apron with one drawer. Double-vase turned legs. Plain square stretchers.

Size of top, 36 x 25 inches

204 HICKORY AND MAPLE FAN-BACK WINDSOR ARMCHAIR

EARLY AMERICAN, ABOUT 1760

Seven spindles piercing the roundabout arm rail. The arms shaped at the ends and supported by turned posts. Saddle seat. Turned and raked legs supported by stretchers with bulbous turnings.

28

HICKORY COMB-BACK WINDSOR ARMCHAIR
NEW JERSEY, ABOUT 1760

[NUMBER 223]

205 MAPLE BUREAU EARLY AMERICAN, ABOUT 1780
Four drawers set in a piped frame. Moulded base; bracket feet.
Height, 34 inches; length, 38 inches

206 MAPLE FOUR-POSTER BED EARLY AMERICAN, ABOUT 1800
Low head and foot board; the head board scrolled. Well turned
posts. The posts on either side of the foot board fluted.
Width, 53 inches

**207 HICKORY AND MAPLE COMB-BACK WINDSOR ARM-
CHAIR** EARLY AMERICAN, ABOUT 1760
Finely shaped comb-back supported by seven slender spindles
which pierce the roundabout arm rail, supported at the end by
turned and raked supports. Saddle seat. Turned and raked legs
with turned stretchers.

**208 PAIR OF RARE BALUSTER-BACK HICKORY AND MAPLE
SIDE CHAIRS** EARLY AMERICAN, ABOUT 1730
Deep scalloped top rail of fine outline between two turned posts
with ball finials. Four-baluster back; round turned legs, with
double stretcher at front and sides. Rush seats with wooden corner pieces. (2)

209 MAPLE SWELL-FRONT BUREAU
EARLY AMRICAN, ABOUT 1790
Very graceful piece. Four drawers in beaded frame gradually
increasing in width towards the base. Bracket feet.
Height, 34½ inches; length, 38 inches

210 QUEEN ANNE WALNUT CHAIR EARLY AMERICAN, 1710-15
Open back with well-carved shell at top and violin-shaped splat.
Wide shaped seat. Cabriole legs ending in pad feet. Turned
stretchers. Slip cushion.

211 OCTAGONAL MAHOGANY OCCASIONAL TABLE
EARLY AMERICAN, ABOUT 1770
Well-turned pedestal support. Three graceful legs ending in
snake-head feet.

212 RARE OAK SWELL-FRONT BUREAU

NEW ENGLAND, ABOUT **1790**

A very unusual piece. Four drawers. Scalloped apron. French bracket feet. Height, 36 inches; length, 39 inches

213 MAPLE SPANISH FOOT SIDE CHAIR

EARLY AMERICAN, ABOUT **1710**

Open back with violin-shaped splat. Rush seat and wooden corner pieces. Elaborately turned front legs with stretcher having double ball turning in front. Spanish feet.

214 HEPPELWHITE INLAID MAHOGANY FOLDING CARD TABLE

EARLY AMERICAN, ABOUT **1790**

Swell front, with satinwood oval panel on rosewood rectangular panel. Front and legs with very delicate line inlay.

Size of top, open, 35¾ x 35 inches

215 CHERRY SWELL-FRONT BUREAU

EARLY AMERICAN, ABOUT **1790**

The four drawers with beaded edge. Deeply scalloped apron inlaid with a narrow mahogany band. French bracket feet.

Length, 40 inches

216 MAPLE HIGH CHEST ON FRAME

EARLY AMERICAN, ABOUT **1720**

Moulded cornice. Six drawers increasing in width towards the base of slightly curly maple. Low Queen Anne legs ending in club feet. Height, 62 inches; length, 39 inches

217 SMALL APPLEWOOD TABLE   EARLY AMERICAN, ABOUT **1720**

Very unusual specimen. Apron with one drawer. The four legs with elaborate turning and connected by central stretcher. The four legs showing a variety of the "turned cup shape" generally found in the early highboys. The legs connected by turned stretchers and brace. Size of top, 29 x 19½ inches

218 MAHOGANY CURLY MAPLE AND SATINWOOD SHERATON DRESSING TABLE

EARLY AMERICAN, ABOUT **1800**

Curly maple top; the front built up of mahogany with two wide satinwood drawers with piping. Turned posts, inlaid with mahogany panels. Height, 32 inches; length, 31 inches

IMPORTANT QUEEN ANNE MAPLE HIGHBOY
EARLY AMERICAN, ABOUT 1710–20

[NUMBER 226]

219 PINE SETTLE
EARLY AMERICAN, 1720-30

The low back with five raised panels; sloped arms; plain end supports. Unusual early piece.
Length, 72½ inches

220 MAPLE CHEST OF DRAWERS
EARLY AMERICAN, ABOUT 1770

Two small drawers above four large drawers. Scalloped and bracketed base. Fine early piece.
Height, 49 inches; length, 37½ inches

221 QUEEN ANNE MAPLE DROP-LEAF TABLE
EARLY AMERICAN, ABOUT 1720

Deeply rounded ends. Cabriole legs ending in club feet. Fine piece. The apron with scalloped ends.
Size of top, open, 42 x 42 inches

222 MAHOGANY DUNCAN PHYFE DROP-LEAF TABLE
EARLY AMERICAN, ABOUT 1810

The drop leaves gracefully rounded. Apron with one drawer. Fluted tapering legs.
Size of top, extended, 46½ x 36 inches

223 HICKORY COMB-BACK WINDSOR ARMCHAIR
NEW JERSEY, ABOUT 1760

The finest type of New Jersey windsor chair, with well-shaped top rail, scrolled at the ends. Roundabout arm rail, pierced by the nine spindles. The arms knuckled at the ends. Saddle seat. Turned and raked legs connected by bulbous turned stretchers.

[SEE ILLUSTRATION]

224 PINE CORNER CABINET IN TWO SECTIONS
EARLY AMERICAN, ABOUT 1780

Elaborately moulded top with dentelled edge. Latticed glass door enclosing three shelves fastened with the original H-shaped hinges. The lower body with two small drawers above two panelled doors enclosing cupboard with one shelf. Ogee bracket feet.
Height, 86 inches; length about 45 inches

225 SET OF FIVE MAHOGANY DUNCAN PHYFE SIDE
CHAIRS   EARLY AMERICAN, ABOUT 1810
Plain panel top rail; horizontal splat finely carved with acanthus
leaves on either side of central rosette. Finely curved sides
and legs. Slip cushions. (5)

226 IMPORTANT QUEEN ANNE MAPLE HIGHBOY
                              EARLY AMERICAN, ABOUT 1710-20
Moulded cornice. Upper body with two small drawers above
three large drawers. Sides in pine. The lower body, with
moulded edge, into which the upper body sets, has three drawers and very deeply scalloped beaded apron. Six legs with fine
cup turnings and deeply scalloped and undulated bracing, on all
sides. Very important piece. Height, 61 inches; length, 36 inches
[SEE ILLUSTRATION]

227 SET OF SIX MAHOGANY HEPPELWHITE SIDE CHAIRS
                              EARLY AMERICAN, ABOUT 1790
Arched top rail with openwork splat, having honeysuckle carving.
Moulded edge. Leather upholstered seat. Fluted square legs.
(6)

228 HEPPELWHITE INLAID WALNUT AND MAHOGANY
SIDEBOARD   EARLY AMERICAN, ABOUT 1790
Very unusual piece, the body in walnut with wide borders of
mahogany around the doors and drawers. Row of three drawers
across the top; below, a deep square drawer at each end and a
cupboard enclosed by two doors in the centre. Six inlaid tapering
legs. Interesting piece.   Height, 42 inches; length, 55 inches

229 HEPPELWHITE INLAID MAHOGANY DROP-LEAF
TABLE WITH EAGLE INLAY   EARLY AMERICAN, ABOUT 1790
Very interesting piece. With oval inlay over each of the legs
with eagle bearing a banderole with five stars in his mouth and the
body formed of a shield; his claws hold an arrow and a sheaf of
wheat. The legs, drawer and top with satinwood linear inlay.
Top with rounded drop leaves. Legs with pendent floral inlay
at the top. Rare piece.   Size of top, 30½ x 30½ inches

INTERESTING HEPPELWHITE INLAID MAHOGANY SECRETARY-DESK
EARLY AMERICAN, ABOUT 1790

[NUMBER 230]

230 INTERESTING HEPPELWHITE INLAID MAHOGANY
SECRETARY-DESK   EARLY AMERICAN, ABOUT 1790
The upper part with cornice having fine shell inlay, brass finials
and phœnix surmounting centre motif. Two latticed glass doors,
with line inlay, enclosing three shelves. Row of three lockers
below, with pigeonholes and small drawers. Fold-back writing
flap. Four drawers below, increasing in depth towards the
base. The base scalloped and with French feet. Old brasses.
Very interesting specimen of old New England furniture, in the
condition in which it was acquired at Newburyport, Mass.

Height, 81 inches; length, 41½ inches
[SEE ILLUSTRATION]

231 FINE MAPLE HIGHBOY   EARLY AMERICAN, ABOUT 1760
The upper body with moulded cornice and five drawers below,
fits into the base, which has a long drawer across the top and
three square drawers below, the centre one with very fine and
deeply carved sunburst motif. Scalloped apron. Cabriole legs
ending in club feet.   Height, 76 inches; length, 38 inches

232 FINE CURLY MAPLE SIDEBOARD
EARLY AMERICAN, ABOUT 1790
Simple rectangular lines. Apron with long drawer in centre,
flanked by two small drawers for silverware. Lower body with
large cupboard enclosed by two panelled doors. Scalloped apron
and French bracket feet. Very beautifully grained wood.

Height, 42 inches; length, 46 inches

233 IMPORTANT DUNCAN PHYFE THREE-PART MAHOG-
ANY DINING ROOM TABLE   EARLY AMERICAN, ABOUT 1810
The three sections each with a square support, turned pedestal
and four plain outcurved legs in brass casings. Very unusual
piece. Seats sixteen people when open.

Length, 7 feet 8 inches; width, 5 feet

234 FINE INLAID MAHOGANY HEPPELWHITE BUREAU
DESK   EARLY AMERICAN, ABOUT 1790
The upper body enclosed by an inlaid drop-leaf panel which
draws out and reveals an interior with finely inlaid drawers and
pigeonholes and a central locker. The three drawers below with
fan inlay in the corners, are outlined in satinwood and ebony.
Scalloped apron. Slender French bracket feet.

Height, 44½ inches; length, 36½ inches

235 PINE AND MAPLE DESK   EARLY AMERICAN, 1700-10

Rare early specimen. The drop lid enclosing deep interior elaborately fitted with three rows of drawers and pigeonholes. Lower body with long drawer. Turned legs connected by plain bracing.

Height, 37 inches; length, 34 inches

236 FINE APPLEWOOD SECRETARY-DESK

EARLY AMERICAN, ABOUT 1775

The upper body with two sunken panel doors, finely arched at the top, enclosing two shelves. Moulded cornice top. Original H-shaped hinges. The lower body with fall front, enclosing interior with six pigeonholes, six drawers and central locker. Four long drawers gradually increasing in depth toward the base. Ogee bracket feet. Fine piece.

Height, 78 inches; length, 39½ inches

237 FINE WALNUT FALL-FRONT WRITING DESK

EARLY AMERICAN, ABOUT 1775

The fall front with very unusual raised and scalloped border. Fine block front interior with four pigeonholes and four small drawers on either side of the central shell-carved door, which encloses secret compartments and three small drawers. Lower body with five drawers. Moulded base and bracket feet.

Height, 42 inches; length, 36½ inches

238 MAPLE FALL-FRONT WRITING DESK

EARLY AMERICAN, ABOUT 1780

The fall front enclosing unusually elaborate system of pigeonholes, small drawers with block front secret compartments, hidden behind pilasters, and central sunburst-carved drawer. Lower body with four wide drawers increasing in depth towards the base. The base supported by bracket feet and with unusual scalloped ornament pendent from the centre.

Height, 41 inches; length, 39½ inches

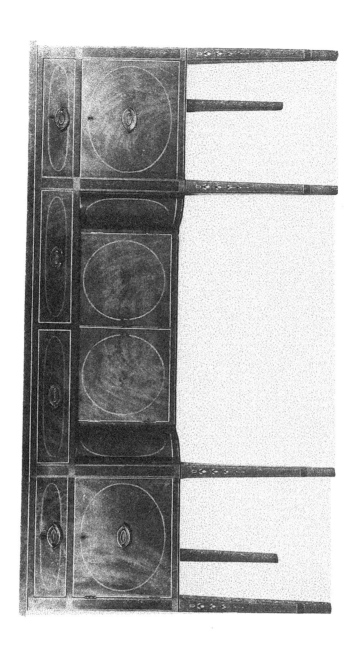

IMPORTANT HEPPELWHITE INLAID MAHOGANY SIDEBOARD
EARLY AMERICAN, ABOUT 1790

[NUMBER 239]

239 IMPORTANT HEPPELWHITE INLAID MAHOGANY SIDE-
BOARD   EARLY AMERICAN, ABOUT 1790

Very fine proportion and elegant simplicity of the inlay. The
plain top edged with a band of satinwood inlay at either side,
and slightly recessed at either end. The apron, which has a long
drawer in the centre flanked by a smaller drawer at each end, is
decorated with four fine oval inlays in satinwood and ebony.
The centre of the lower body deeply recessed and with two doors
having circular inlay flanked by convex panels with oval inlay.
At the right end a deep drawer and to the left a cabinet enclosed
by a door, both with round inlay. Square, slightly tapering legs
with pendent leaf and linear inlay and ebony banding above the
feet. Ebony key plates. Very fine specimen.

Length, 70 inches; height, 40¼ inches

[SEE ILLUSTRATION]

240 HEPPELWHITE   INLAID   MAHOGANY   AND   CURLY
MAPLE WRITING DESK   EARLY AMERICAN, ABOUT 1790

The recessed top section enclosed by three doors framed by a
wide rosewood banding piped in satinwood; the two end doors
each open on two drawers and three pigeonholes; the narrow
centre door encloses deep pigeonhole and drawer. In front of
the pigeonhole section, fold-back writing section supported by
pulls; the edge of the flap with wide rosewood inlay. Lower
body with four curly maple drawers framed in rosewood. Deeply
scalloped apron and French bracket feet.

Height, 46½ inches; length, 40 inches

241 CHIPPENDALE MAHOGANY WALL MIRROR

EARLY AMERICAN, ABOUT 1770

Fine piece with finely scrolled top having a gilt phœnix ornament
in the centre. Gilt acanthus ornaments at the sides. Inner
moulding scalloped and gilt. Gilt acanthus side brackets.

Size, 7½ x 23 inches

242 SET OF TEN CHIPPENDALE MAHOGANY SIDE CHAIRS

EARLY AMERICAN, ABOUT 1780

Very fine and unusual set. Bow-shaped top rail, with pierced,
violin-shaped splat, having twisted ribbon motif in the centre.
Cabriole front legs, ending in pad feet. Slip seats. (10)
See a similar specimen in Lockwood, page 86, Fig. 543.

[SEE ILLUSTRATION]

243 BONNET-TOP CHERRY HIGHBOY

EARLY AMERICAN, ABOUT **1740**

The broken arched top with urn finial at each end and in centre. Below the central finial, drawer with finely carved sunburst flanked by two smaller drawers. Four long drawers below increasing in width towards the base. Lower body with two drawers; the lower one panelled to simulate three drawers with sunburst carving in the centre. Cabriole legs with club feet.

Height, 83 inches; length, 36½ inches

244 HEPPELWHITE INLAID MAHOGANY SIDEBOARD

EARLY AMERICAN, ABOUT **1790**

The top, which is outlined by satinwood banding, is outcurved at either end and straight across the front. Each of the outcurved ends with two shelves enclosed by a curved front door inlaid in satinwood to simulate a narrow and a deep drawer. The centre of the sideboard with a narrow drawer at the top and below, a deep cupboard enclosed by two inlaid doors flanked by narrow inlaid wine drawers. Square tapering legs with pendant inlay. Very fine specimen.   Height, 42 inches; length, 68 inches

245 EXTREMELY RARE MAPLE DAY-BED

EARLY AMERICAN, ABOUT **1680**

Flemish style. Turned posts, supporting the movable back, with arched top, and openwork scroll carving. Back and seat with caning of later date. The seat supported by eight bracket feet connected by arched stretchers. Very rare specimen, which may be compared with Fig. 637 in Lockwood.   Length, 61 inches

[SEE ILLUSTRATION]

246 MAPLE HIGHBOY

EARLY AMERICAN, ABOUT **1720**

Moulded top. Upper body with row of three small drawers below the cornice. The centre drawer with fine sunburst carving. Four long drawers below, increasing in depth towards the base. Lower body with three square drawers, the centre with deep and finely fluted sunburst carving. Scalloped apron. Cabriole legs ending in club feet.   Height, 69½ inches; length, 39½ inches

SET OF TEN CHIPPENDALE MAHOGANY SIDE CHAIRS
EARLY AMERICAN, ABOUT 1780

[NUMBER 242]

247 SET OF FOUR MAHOGANY SHERATON SIDE CHAIRS
EARLY AMERICAN, ABOUT 1790

The square open back with narrow frieze of laurel carving in centre and two fluted slats carved with palm leaves at the top. Horsehair seats studded with nails. Square legs connected by braces. A fine set. (4)

OO

248 CHIPPENDALE CHERRYWOOD LOWBOY
EARLY AMERICAN, ABOUT 1770

Apron with one low drawer, three square drawers below, the centre one with fine sunburst carving. Deeply scrolled and scalloped apron. Cabriole legs with claw and ball feet. Original brasses.

Height, 30 inches; length, 36 inches.

10

249 FINE BURLED FRONT CHEST OF DRAWERS
EARLY AMERICAN, ABOUT 1700

Pine top, curly maple sides. Two small and three large drawers of very beautiful grained wood, framed by a double walnut herringbone frieze. Moulded base with round walnut feet. Original teardrop handles.   Height, 34 inches; length, 35 inches

250 MAPLE HIGHBOY   EARLY AMERICAN, ABOUT 1720

Unusually graceful piece. Moulded top with five drawers below, the upper row of two small drawers. The top sets into moulded base with wide drawer panelled in front to simulate three square drawers. Scalloped apron. Very slender cabriole legs ending in pad feet.   Height, 64½ inches; length, 39½ inches

251 MAPLE SWELL-FRONT BUREAU
EARLY AMERICAN, ABOUT 1790

The drawers set into piped moulding. Moulded base with French bracket feet. Very graceful piece.

Height, 35 inches; length, 40 inches

252 MAPLE FALL-FRONT WRITING DESK
EARLY AMERICAN, ABOUT 1780

Fall front, enclosing one row of pigeonholes and two rows of small drawers, and supported by two pulls. Lower body with four wide drawers, increasing in depth towards the base. Bracket feet.   Height, 43 inches; length, 26 inches

253 MAPLE AND HICKORY ARMCHAIR IN DUTCH STYLE

EARLY AMERICAN, ABOUT **1740**

Arched top rail supported by turned posts and with well-shaped baluster splat. Straight arms scrolled at the ends and supported by the turned legs. Double turned bracing at the front; plain bracing at the sides. Rush seat.

254 MAHOGANY OCCASIONAL TABLE WITH TWO DRAWERS

EARLY AMERICAN, ABOUT **1790**

Square top with reeded edge. Two drawers with beaded moulding. Slender square legs connected by X-shaped brace.

Size of top, 17 x 16½ inches

255 RARE MAPLE CANDLE STAND

EARLY AMERICAN, ABOUT **1725**

Adjustable dish top on the spirally turned pedestal support. Adjustable candle holder fitted for two lights above. Three outcurved legs.

256 PINE AND MAPLE TAVERN TABLE

EARLY AMERICAN, ABOUT **1715**

Pine top; maple base. Apron fitted with drawer. Slender turned legs connected by plain bracing. Size of top, 23 x 35 inches

257 PINE AND MAPLE TAVERN TABLE

EARLY AMERICAN, ABOUT **1740**

Pine top. Turned legs connected by plain stretchers.

Size of top, 34 x 25 inches

258 PINE AND MAPLE TAVERN TABLE

EARLY AMERICAN, ABOUT **1740**

Plain pine top; turned legs, connected by plain stretchers.

Size of top, 33 x 22 inches

259 MAHOGANY TIP-TOP TABLE   EARLY AMERICAN, ABOUT 1800

Top with deeply scalloped corners. Urn and baluster support. Three outcurved legs. Size of top, 21 x 17 inches

260 CURLY MAPLE AND CHERRY TIP-TOP TABLE

EARLY AMERICAN, ABOUT **1800**

The top with finely grained cherrywood has cut and scalloped corners. Turned and ringed urn-shaped pedestal support in curly maple. Three outcurved cherry legs. Size of top, 19 x 17 inches

38

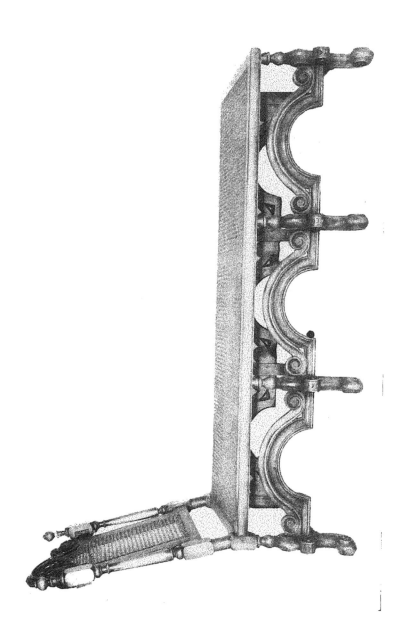

EXTREMELY RARE MAPLE DAY-BED
EARLY AMERICAN, ABOUT 1680

[NUMBER 245]

261 MINIATURE PINE CHEST Early American, about 1690

Drop lid with moulded edge. One drawer at base. Wooden knobs. Round turned feet.

Length, 17½ inches

262 MAHOGANY KIDNEY-SHAPED TIP-TOP TABLE
Early American, about 1790

Gracefully shaped pedestal support with three outcurved bracket legs.

Length, 23 inches

263 HEPPELWHITE INLAID MAHOGANY SHAVING STAND
Early American, 1790

The square mirror, with inlaid satinwood edge, is supported between plain turned posts. Swell front with three drawers, edged in satinwood.

264 SQUARE MAPLE CANDLE STAND
Early American, about 1770

Very slender urn-shaped baluster support with three finely shaped legs of elegant design. Unusually graceful piece.

265 APPLEWOOD TIP-TOP TABLE Early American, about 1760

Round top with turned vase-shaped columnar support and three legs ending in snake-head feet.

266 HICKORY AND MAPLE COMB-BACK SIDE CHAIR
Early American, about 1760

Seven slats; slender turned posts. Saddle seat. Turned and raked legs connected by bulbous turned stretcher. Graceful specimen.

This catalogue designed by The Anderson Galleries
Composition and Press-work by
B. H. Tyrrel, New York

Lightning Source UK Ltd.
Milton Keynes UK
UKHW031809150119
335176UK00013BA/1768/P